AF228569

SEA OTTERS

Kelp Forest Keepers

MEGAN BORGERT-SPANIOL

Consulting Editor, Diane Craig, M.A./Reading Specialist

Super Sandcastle

An Imprint of Abdo Publishing
abdobooks.com

ABDOBOOKS.COM

Published by Abdo Publishing, a division of ABDO, PO Box 398166, Minneapolis, Minnesota 55439.

Printed in the United States of America, North Mankato, Minnesota
102019
012020

THIS BOOK CONTAINS
RECYCLED MATERIALS

Design: Kelly Doudna, Mighty Media, Inc.
Production: Mighty Media, Inc.
Editor: Liz Salzmann
Cover Photographs: iStockphoto, Shutterstock Images
Interior Photographs: Getty Images/iStockphoto, pp. 3, 4, 6, 7, 10, 11, 13, 16, 17, 18; Shutterstock
Images, pp. 4, 5, 6, 7, 8, 9, 11, 12, 13, 14, 15, 16, 17, 18, 19, 20, 21, 22, 23

Publisher's Cataloging-in-Publication Data
Names: Borgert-Spaniol, Megan, author.
Title: Sea otters: kelp forest keepers / by Megan Borgert-Spaniol
Other title: kelp forest keepers
Description: Minneapolis, Minnesota : Abdo Publishing, 2020 | Series: Animal eco influencers
Identifiers: ISBN 9781532191893 (lib. bdg.) | ISBN 9781532178627 (ebook)
Subjects: LCSH: Sea otter--Juvenile literature. | Sea otter--Behavior--Juvenile literature. | Kelp forest
    ecology--Juvenile literature. | Ocean animals--Juvenile literature. | Animal ecology--Juvenile
    literature. | Wildlife habitats--Juvenile literature.
Classification: DDC 599.74447--dc23

CONTENTS

ECO INFLUENCERS

What is an eco **influencer**? It is an animal that can change its **ecosystem**. All members of an ecosystem affect one another.

Beavers build structures that create **wetlands**. The wetlands support other wildlife.

Gray wolves hunt elk in the forest. This means elk eat fewer plants. So, there are more plants for other wildlife to eat.

Sea otters are eco **influencers**. They maintain **kelp** forests by eating sea urchins. Sea otters are kelp forest keepers!

PACIFIC SWIMMERS

Sea otters live in the Pacific Ocean. They swim near the coasts of **North America** and **Asia**.

Giant kelp can be more than 150 feet (45 m) tall.

KELP FORESTS

Sea otters live in and around **kelp** forests. Kelp is a brown, leafy seaweed. It grows in thick groups called kelp forests. These forests grow in shallow coastal waters.

About 90 percent of the world's sea otters live near Alaska's coasts.

FUR COATS

Sea otters spend most of their lives in the cold ocean water. But their fur keeps them dry and warm.

Sea otter fur is **waterproof**. This keeps the animal dry. The fur also traps air next to the otter's skin. This keeps the otter warm.

Sea otters have the thickest fur in the animal kingdom. Their coats have up to one million hairs per square inch!

THINK!
How does the hair
on your body react to the
cold? Next time you feel cold,
look at the hairs on your
arm. They will stand on end
to trap heat, just like
an otter's fur!

9

GROOMING AND FLOATING

For sea otters to keep warm and dry, their fur must stay clean. Sea otters spend a lot of time **grooming** their fur. They use their teeth and paws to clean their coats.

Sea otters do most activities while floating on their backs! This includes eating, grooming, sleeping, and carrying their young.

THINK!
A sea otter's body
is built for floating.
What is your body
built for?

SHELL CRUSHERS

Sea otters eat shellfish. These include mussels, clams, crabs, and sea urchins. Sea otters dive to the ocean floor to find food. Then they bring their catches back to the surface.

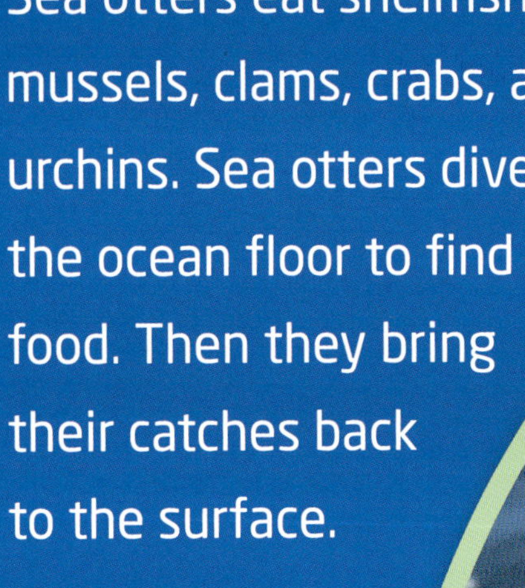

Sea otters have powerful jaws for crushing the shells of their prey.

EATING TOOLS

Sometimes sea otters use rocks to crack open shellfish. An otter sets a rock on its stomach. Then it bangs the shellfish against the rock. This cracks the shell.

13

URCHIN EATERS

Sea otters act as predators in the **food web**. This is one way they **influence** their **ecosystem**.

Sea urchins are one of the main foods sea otters eat. These spiny shellfish eat **kelp**. Without predators, sea urchins would wipe out kelp forests.

A kelp
forest can
die if sea urchins
take over.
Sea
urchins
eat kelp.
An area
taken over
by sea urchins is
called an urchin
barren.

KELP FORESTS

Sea otters keep sea urchins from taking over **kelp** forests. This helps the kelp stay healthy. Kelp forests are important to the coastal **ecosystem**. They provide **habitats** for many ocean animals!

Fish, snails, sea stars, and other small creatures find food and shelter in kelp forests.

Gulls
and other
seabirds eat
small creatures
living among
the kelp.

Seals,
whales, and
other animals hide
from predators in kelp
forests. Kelp forests
also provide shelter
during storms.

HEALTHY OCEANS

Kelp forests do more than provide food and shelter. They also capture **carbon dioxide**, or CO_2.

Human activities have raised CO_2 levels in the air and ocean. CO_2 makes the ocean warmer and more acidic. These changes are harming ocean life and **ecosystems**.

We produce CO_2 when we burn fuels for energy.

Kelp uses CO_2 for energy. This removes CO_2 from the ocean water. By maintaining kelp forests, sea otters help maintain the ocean **ecosystem**!

A healthy kelp forest

THINK!

What are some ways your ecosystem keeps you alive and healthy?

SAVING SEA OTTERS

In the 1700s and 1800s, sea otters were hunted for their fur. They almost died out. Today, laws protect sea otters from being hunted.

But sea otters are still in danger. Certain types of fishing can harm sea otters. So can pollution.

Many people are working to save sea otters. The ocean needs these **kelp** forest keepers!

Sea otters can get caught in shellfish traps.

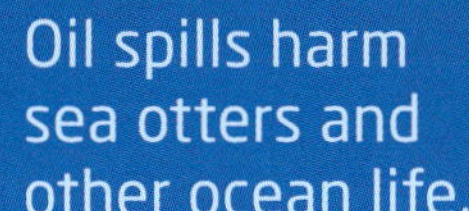

Oil spills harm sea otters and other ocean life.

FUN FACTS

Sea otters can dive more than 300 feet (90 m) deep when looking for food.

Sea otters can eat about 25 percent of their weight in food each day.

Sea otters place **kelp** over their bodies when they rest. This keeps them from drifting.

SEA OTTER QUIZ

1. In what ocean do sea otters live?

2. Sea otters eat **kelp**.

    True or **false**?

3. What does kelp use for energy?

       A. oxygen

       B. carbon dioxide (CO_2)

       C. water

Answers: 1. Pacific Ocean 2. False 3. B

GLOSSARY

Asia—the largest of the continents. Russia, India, and China are in Asia.

carbon dioxide (CO$_2$)—a gas mainly produced in the bodies of people and animals, which becomes part of the air they breathe out.

ecosystem—a group of plants and animals that live together in nature and depend on each other to survive.

food web—the feeding relationships between different organisms in a community.

groom—to clean and care for.

gull—a mostly white or gray bird with long wings and webbed feet that usually lives near water.

habitat—the area or environment where a person or animal usually lives.

influence—to cause something to change.

kelp—a large, brown seaweed.

North America—the continent that includes Canada, the United States, and Mexico as well as other countries.

react—to move or behave in a certain way because of something else.

waterproof—made so that water can't get in.

wetland—a low, wet area of land such as a swamp or a marsh.